FALCONS

by Jaclyn Jaycox

PEBBLE
a capstone imprint

Published by Pebble, an imprint of Capstone
1710 Roe Crest Drive, North Mankato, Minnesota 56003
capstonepub.com

Library of Congress Cataloging-in-Publication Data
Names: Jaycox, Jaclyn, 1983- author.
Title: Falcons / by Jaclyn Jaycox.
Description: North Mankato, Minnesota : Pebble, [2023] | Series: Animals | Includes bibliographical references and index. | Audience: Ages 5-8 | Audience: Grades K-1 | Summary: "Falcons dive through the air. These birds of prey can reach speeds of more than 240 miles an hour! Their thin bodies and long, pointed wings help them go fast. Young readers will be quick to learn all they can about fierce falcons with fun facts and exciting photos"— Provided by publisher.
Identifiers: LCCN 2022000672 (print) | LCCN 2022000673 (ebook) | ISBN 9781666342758 (hardcover) | ISBN 9781666342796 (paperback) | ISBN 9781666342833 (pdf) | ISBN 9781666342918 (kindle edition)
Subjects: LCSH: Falcons—Juvenile literature.
Classification: LCC QL696.F34 J39 2023 (print) | LCC QL696.F34 (ebook) | DDC 598.9/6—dc23/eng/20220202
LC record available at https://lccn.loc.gov/2022000672
LC ebook record available at https://lccn.loc.gov/2022000673

Image Credits
Capstone Press, 6; Getty Images: Galen Rowell, 25, Mark Newman/Design Pics, 7; Shutterstock: aaltair, 19, Adam Morse 2, 1, Agami Photo Agency, 10, Alex Alderic Jero, 20, Ali Bernie Buga-ay, Cover, Bouke Atema, 9, Chris Hill, 14, Dagmara Ksandrova, 15, EcoPrint, 28, Giedriius, 26, Harry Collins Photography, 11, 17, Ken Griffiths, 21, L-N, 22, Megan M. Weber, 12, PongMoji, 27, Rob Palmer Photography, 13, Sriram Bird Photographer, 5, TPCImagery - Mike Jackson, 18

Editorial Credits
Editor: Abby Huff; Designer: Dina Her; Media Researchers: Jo Miller and Pam Mitsakos; Production Specialist: Tori Abraham

All internet sites appearing in back matter were available and accurate when this book was sent to press.

Table of Contents

Words in **bold** are in the glossary.

Amazing Falcons

What kind of bird is faster than a race car? A falcon! It's not only quick. It's fierce too! These animals are known for being great hunters.

Falcons are a type of **raptor**. These birds hunt and eat small animals. They are also called birds of **prey**. There are about 40 kinds of falcons.

Where in the World

Falcons are found all around the world. They live on every **continent** except Antarctica.

Falcons Range Map

North America

Europe

Asia

Pacific Ocean

Atlantic Ocean

Pacific Ocean

Africa

Range

N
W E
S

South America

Indian Ocean

Australia

Southern Ocean

Antarctica

Falcons live in many **habitats**. Some are found where it's very cold. Others live in warm areas. They live in grasslands, mountains, or forests. They can live in the icy **tundra** or hot deserts too.

Some kinds of falcons **migrate**. They fly to warmer areas during the winter. Then they return in the spring. These trips can be long. Some falcons fly 15,500 miles (25,000 kilometers) each year. That's like traveling across the U.S. more than five times!

Most falcons live high off the ground. They sit in trees or on cliffs. Some even live in busy cities. They make nests on skyscrapers or tall bridges.

Falcon Bodies

A falcon's body is made for speed. It has a small head and thin body. It has long, pointed wings. Falcons reach their top speeds while diving through the air.

A diving peregrine falcon

Peregrine falcons are the fastest. They dive three times faster than a cheetah can run! They reach speeds of more than 240 miles (386 km) per hour. No other animal on Earth can move that fast.

American kestrel

Falcons can be many colors. Some have gray, brown, or black feathers. Others are yellow, red, or white. Males and females are usually the same colors.

American kestrels are some of the smallest falcons. They grow 8 to 12 inches (20.3 to 30.5 centimeters) long. Gyrfalcons are the largest. They grow up to about 2 feet (0.6 meter) long. Their wingspans are twice as long as their bodies.

gyrfalcon

Falcons have large claws called **talons**. Many raptors use their talons to kill prey. But not falcons. They may use their feet to grab an animal or knock it down. But their beak is their secret weapon.

talon

A falcon's strong beak has a hook.
The hook acts like a tooth. It can easily
tear through prey and break bones.

On the Menu

A hungry falcon sits high up in a tree. It spots a bird flying below. The falcon dives. *Zoom!* The bird tries to get away. But it's no match for a speedy falcon. The falcon grabs the bird in its talons. Dinnertime!

Falcons eat meat. They often eat other birds. Some falcons hunt bats, rabbits, and mice. They eat **reptiles** and insects too.

A peregrine falcon takes off to grab its prey.

Falcons usually hunt during the day.
They have great eyesight. They can
spot prey almost 2 miles (3 km) away.

Falcons often snatch prey as they fly. Sometimes they land to eat. But they can also eat while flying! They hold the prey in their claws. They reach down to take bites while moving through the air.

Life of a Falcon

Falcons live most of their lives alone. They only come together to **mate**. But they mate for life. The same male and female find their way back to each other every year.

When a female falcon makes a nest, she uses her feet. She digs out a small spot in sand or dirt. She makes a bowl-like area for eggs. This type of nest is called a scrape. Falcons will also use nests left by other birds.

A female usually lays three to four eggs at a time. She sits on the eggs. This keeps them warm. The male will sometimes sit on the eggs too. But he mostly hunts. He brings the female food. After a few weeks, the eggs hatch.

Chicks are born with fuzzy feathers. Both parents help feed the young. After about a month, chicks lose their baby feathers. Their adult feathers grow in.

Young falcons grow quickly. They are full-grown by six weeks. They start flying out of the nest. But they don't go far. They come back to the nest to sleep. After a few more weeks, they are ready for life on their own.

Smaller falcons start to mate after one year. Larger falcons may not mate for three years. Falcons can live 12 to 20 years.

Young prairie falcons

Dangers to Falcons

Falcons are strong, fast birds. But they still have **predators**. Wolves hunt them. Eagles and owls will eat falcon chicks.

golden eagle

A person sprays poison to control insects.

Humans have been a danger to falcons. Around 1970, some falcons in North America were in danger of dying out. Humans were using a certain poison to control insects. Birds ate things with the poison on it. Then falcons ate those birds. Many falcons died.

In 1972, the U.S. made new laws. The poison could not be used anymore. The number of falcons went up. Today, people are still working to help falcons.

Falcons' nesting areas are now kept safe by laws. Groups teach people about falcons. They don't want to put these amazing animals at risk again.

Fast Facts

Name: falcon

Habitat: mountains, grasslands, deserts, forests, tundra

Where in the World: every continent except Antarctica

Food: birds, bats, rabbits, mice, reptiles, insects

Predators: wolves, owls, eagles, humans

Life Span: 12 to 20 years

Glossary

continent (KAHN-tuh-nuhnt)—one of Earth's seven large land masses

habitat (HAB-uh-tat)—the natural place where a plant or animal lives

mate (MEYT)—to join with another to produce young

migrate (MAHY-greyt)—to move from one place to another at different times in the year

predator (PREH-duh-tur)—an animal that hunts other animals for food

prey (PRAY)—an animal hunted by another animal for food

raptor (RAP-tohr)—a bird that hunts and eats small animals; also called a bird of prey

reptile (REP-tile)—a cold-blooded animal that breathes air, has a backbone, and often has scales

talon (TAL-uhn)—a long, sharp, hooked claw

tundra (TUHN-druh)—a cold area where trees do not grow and the soil underground is always frozen

Read More

Riggs, Kate. *Falcons*. Mankato, MN: The Creative Company, 2023.

Sommer, Nathan. *Falcons*. Minneapolis: Bellwether Media, 2019.

Walker, Alan. *Birds of Prey*. New York: Crabtree Publishing, 2022.

Internet Sites

DK FindOut!: Falcons
dkfindout.com/us/animals-and-nature/birds/falcons/

National Geographic Kids: Peregrine Falcon
kids.nationalgeographic.com/animals/birds/facts/peregrine-falcon

Science Trek: Birds of Prey Video Short
pbslearningmedia.org/resource/idptv11.sci.life.oate.d4kbop/birds-of-prey/

Index

About the Author

Jaclyn Jaycox is a children's book author and editor. She lives in southern Minnesota with her husband, two kids, and a spunky goldendoodle.